Food 223

花的力量

Flower Power

Gunter Pauli

[比]冈特·鲍利 著

[哥伦]凯瑟琳娜·巴赫 绘

颜莹莹 译

上海远东出版社

特别感谢以下热心人士对童书工作的支持：

匡志强　方　芳　宋小华　解　东　厉　云　李　婧

刘　丹　熊彩虹　罗淑怡　旷　婉　杨　荣　刘学振

何圣霖　王必斗　潘林平　熊志强　廖清州　谭燕宁

王　征　白　纯　张林霞　寿颖慧　罗　佳　傅　俊

胡海朋　白永喆　韦小宏　李　杰　欧　亮

目录

Contents

一朵玉兰和一朵非洲菊在赞美花的力量。他们在回顾过去这么多年里，他们这两个物种是如何出现，并呈现出惊艳而多样的颜色和形状的。非洲菊说：

“能给世界带来这么多美丽和色彩，是多么快乐啊。我们每天都给人以灵感。我们出现在他们的花园里、家里、他们的艺术品和照片里……”

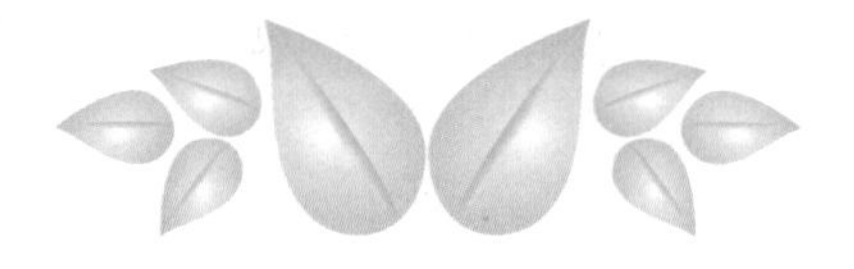

A magnolia and an African daisy are celebrating flower power. They are reflecting on how both their species have emerged, through the ages, with an astounding diversity of colour and shape. The African daisy says:

“What a delight to offer the world so much beauty and colour. We inspire people every day, you know. We are in their gardens, their homes, their art, their photographs...”

一朵玉兰和一朵非洲菊……

A magnolia and an African daisy ...

非洲菊就像华丽的地毯……

Exuberant carpets of daisies ...

“……还有野外，非洲菊就像华丽的地毯一直铺满了南非西海岸。你们这些非洲菊，真的展现出花儿才是世上独一无二的！”

“现在你只是在炫耀，使用‘华丽’这样优雅的词。嗯，也许我也有点爱炫耀。”

“你这是什么意思？”玉兰问道。

“嗯，我喜欢炫耀我的美貌，让人们爱慕我。”

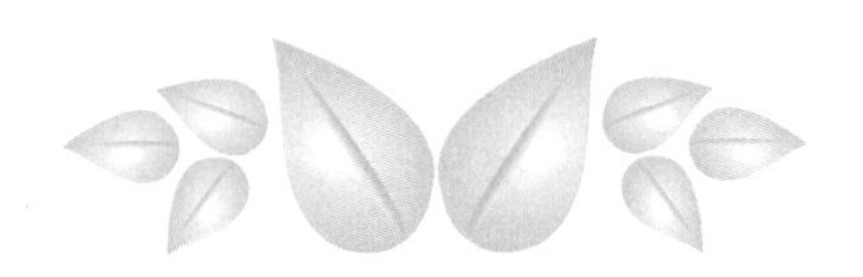

“… And in the wild, like those exuberant carpets of daisies that stretch all along the West Coast of South Africa. You daisies really put on a display that celebrates flowers like no one else in the world!”

“Now you’re just showing off, using big words like ‘exuberant’. Well, maybe I am a bit of a show-off too.”

“What do you mean?” Magnolia asks.

“Well, I like showing off my beauty, and having people adore me.”

“不光是你。我们都喜欢人们靠近观赏，并为我们的精巧美丽所折服。他们一定在第一次花开时就是这样。在日本，当樱花盛开时，整个国家都欣喜若狂。”

“你知道吗？起初，我们的星球上到处都是松树。和开花植物相比，那看上去一定很无聊。松果根本不及我们美丽。谁会喜欢它们呢？”

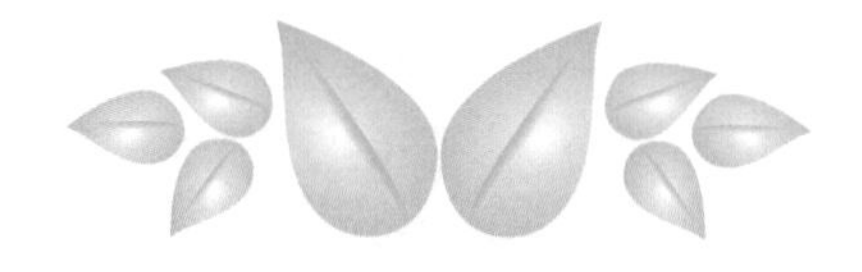

“You are not alone. We all like it when people take a closer look, and are in awe of our beauty and ingenuity. They must have done so since the first flowers popped up. In Japan, an entire nation goes ecstatic when the cherry blossoms peak.”

“Did you know that our planet was, at first, dominated by pine trees? That must have looked very boring compared to the flowering plants. Pinecones cannot possibly compete with our beauty. Who would prefer them?”

……当樱花盛开时。

... when the cherry blossoms peak.

松鼠，我肯定！

The squirrels, I'm sure!

“松鼠，我肯定！我们为自己感到骄傲，但如果没有与我们一起进化的那些昆虫，我们不可能成为最大的植物家族。”

“谢谢你提醒我，玉兰。不是你让蜜蜂开始给花授粉的吗？它们一开始一定笨手笨脚的，想要得到花蜜。”

“它们经验丰富，堪称很好的合作伙伴。看，多亏了它们的辛勤工作，世界上才会有这么多食物。”

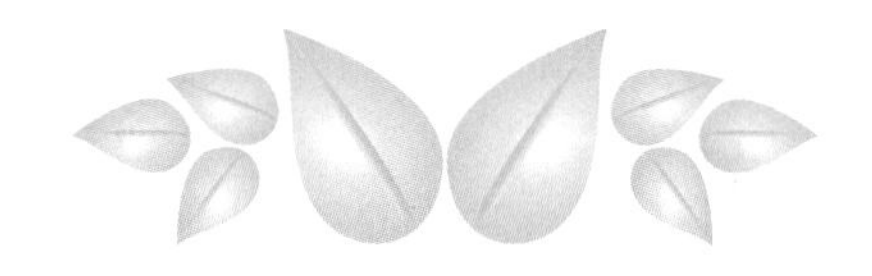

“The squirrels, I’m sure! We are so proud of ourselves, but we could not have become the biggest family of plants had it not been for the insects that have evolved alongside us.”

“Thanks for reminding me of that, Magnolia. Weren’t you the one who got bees to start pollinating flowers? They must have been quite clumsy at first, trying to get to the nectar.”

“Well, they’ve learnt their lessons well, and turned out to be great partners. Look at all the food the world has thanks to their hard work.”

“昆虫的确是很棒的合作伙伴，不是吗？我很喜欢这个创意，我们花儿设计诱使他们给我们授粉，然后再给他们回报！”非洲菊说。

“别忘了，我们需要的不仅仅是传粉者。为了长出令人惊叹的花朵，我们还需要肥沃的土壤来滋养我们的根、真菌、小虫子和虱子，还有……”

“完全正确！人们很容易忽略其实我们有两张生命之网：一张在地面上，太阳是我们的能量来源，另一张是地下的食物网。”

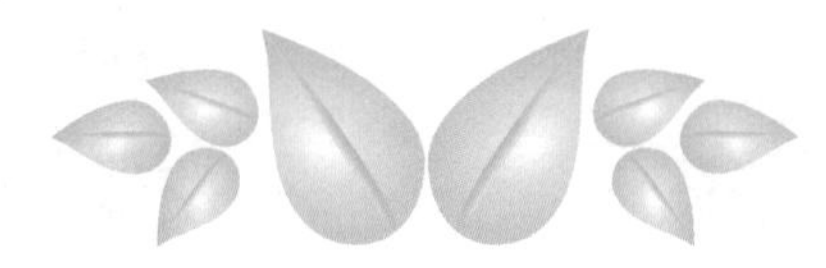

“Insects do make marvellous partners, don’t they? I do enjoy the creative ways we flowers trap, trick and treat them into pollinating us, and them getting rewarded for it!” Daisy says.

“We should not forget that we need more than just pollinators. To grow our awe-inspiring flowers, we also need rich soil to feed our roots, fungi, small bugs, and lice, and …”

“Exactly! People so easily overlook the fact that we have two webs of life: the one above the ground, with the sun as our power source, and then the underground food web.”

……真菌、小虫子和虱子……

… fungi, small bugs, and lice …

在自然界没有任何流失，所有东西都会被循环利用。

In Nature nothing is lost, all is recycled.

“地下没有阳光，那里的能量来源是什么？以什么为食？”

“尸体！难道你不知道我们的脚下很恐怖吗？”

“不要称其为恐怖！死亡是生命的一部分，有机体要想茁壮成长，就必须改变它们所拥有的一切。有机物腐烂后会将营养还给生态系统。记住：在自然界没有任何流失，所有东西都会被循环利用。”

“嗯，我确实让那些地下生物啃我的根。它们能很好地将腐烂残渣转化并形成优质的土壤。它们甚至会吸收空气中大量的碳。”

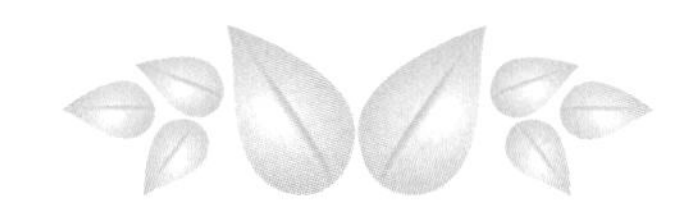

“With no sun underground, what is the power source there? What is being fed on?”

“Dead bodies! Did you not know that there was this creepy reality underneath our feet?”

“Don’t call it creepy! Death is part of life, and for organisms to thrive they need to transform everything they have. Organic matter decays and returns nutrients to the ecosystem. Remember: in Nature nothing is lost, all is recycled.”

“Well, I do let those underground creatures nibble on my roots. They do such a good job converting debris and building up good soil. They even pull lots of carbon out of the air.”

“是的，有了这些地上和地下的伙伴，如果危机来临，我们确实有更好的生存机会。”

“咱们开花植物真聪明！我们找到一个如何让一片普通的绿叶变成一朵美丽、多彩、奇妙的花朵的方法。”

“你的意思是说，我们花朵，起初，只是绿叶？”玉兰问道。

“对。你能想象促成这种改变的设计吗？这就像毛毛虫变成蝴蝶。是的，这是一项重大的创新。”

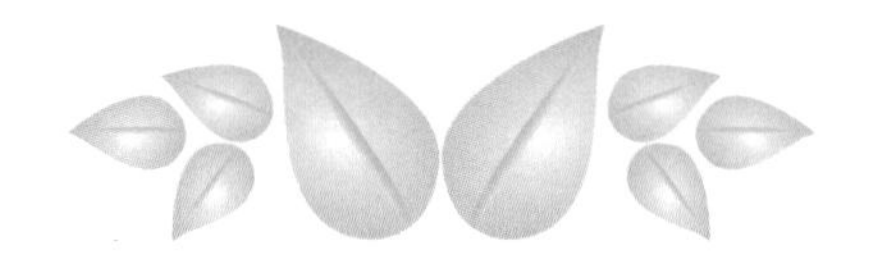

“Yes, with all the partners we have, above and under the ground, we do stand a better chance of surviving should a crisis hit.”

“We flowering plants are really very clever! We figured out how to transform a simple green leaf into a beautiful, colourful, mystical flower.”

“You mean that we flowers were, at first, just green leaves?” Magnolia asks.

“Yip. Can you imagine the design that underpins such a transformation … it’s like a caterpillar turning into a butterfly. An innovation of that magnitude, yes.”

这就像毛毛虫变成蝴蝶。

It's like a caterpillar turning into a butterfly.

向日葵——适应太阳的冠军。

This Sunflower- the champion of adaptation.

“还有更多：对折的叶子适应了每一个授粉伙伴，使我们每一朵花都有独特的姿态。”

“我认为没有哪个家族和我们一样形形色色。没有谁能带来这么多惊喜和快乐——从美丽的外表到多样的颜色、形状，以及香味……”

“是的，举个例子，看看向日葵吧，有上千个品种。这个家族是适应太阳的冠军！”

“那么兰花和豌豆呢？兰花的种子最小，只有尘埃那么大，所以它们可以飞到热带雨林的树冠上去。”

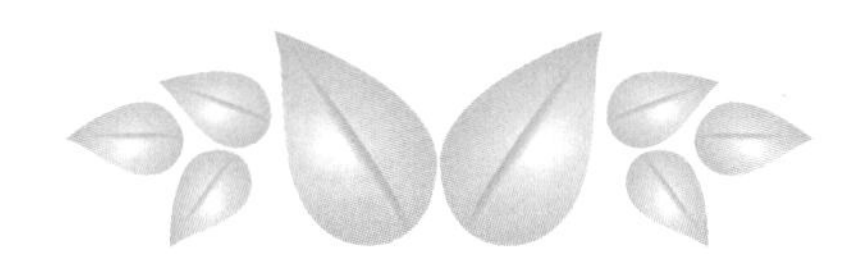

“There is more: a folded leaf then adapted to each pollinating partner, and created a unique niche for each one of us.”

“I don’t think there is any family as diverse as ours. There’s not one that offers so many surprises, or offers so much pleasure – from beauty, to colour, to shape, to fragrance…”

“Yes, just look at the thousands of varieties of the sunflower, for instance. This family is the champion of adaptation, soaking in the sun!”

“And what about the orchids, and the peas? Orchids produce the smallest seeds, the size of a dust particle, so they fly high in the canopy of the rainforest.”

“非洲菊，你知道吗，人们用剪枝工具把我们开花植物的茎剪下来，然后插到水里或土里，几周内根就长出来了。”

“人类是那么做的。我们则运用我们的美貌去吸引和分享，使地球生机勃勃。”

“这正是花的力量！我们是植物世界和动物世界的神奇交汇，生命在这里不断再生！”

……这仅仅是开始！……

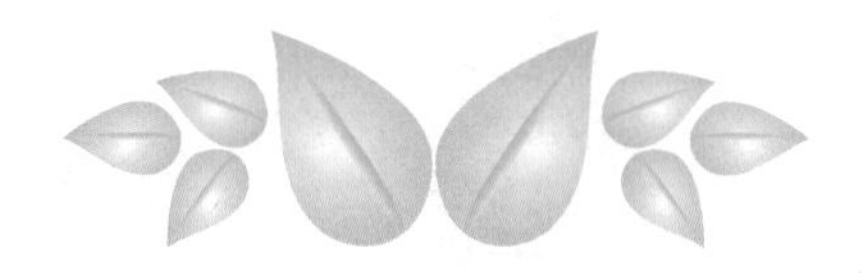

“Daisy, did you know that people use a cutting tool to cut stems then stick them in water or soil, and in a few weeks we flowering plants are growing roots.”

“That's what people do. Our tools are our beauty, seduction and sharing, to promote life on Earth.”

“Flower power, indeed! We are the magical interface between the world of plants and the world of animals, where life regenerates itself – over and over again!”

... AND IT HAS ONLY JUST BEGUN!...

……这仅仅是开始！……

... AND IT HAS ONLY JUST BEGUN! ...

Did You Know ?

你知道吗？

Roughly 350,000 species of flowering plants make up 90% of all plant species. Without them, we would have none of our major crops that feed people and livestock. Flowering plants, along with the soil in which they thrive, form an important carbon sink.

大约有35万种开花植物，占所有植物种类的90%。没有它们，就不会有养活人类和牲畜的那些主要农作物。开花植物和它们所生长的土壤一起构成了一个重要的碳汇。

Flowers did not exist for most of the Earth's history. Early forests consisted of primate plants like club mosses and horsetails. Conifers, known as angiosperms, took over the world of plants before the flowering variety arrived.

在地球历史上，大部分时间里并没有花。早期的森林由诸如石松类和楔叶类等远古植物组成。在开花植物出现之前，植物世界由针叶树这种被子植物占据。

Genome doubling gives extra genetic material that allows for new traits to appear in a plant (or an animal). Genome doubling gave plants the ability to evolve new, never before seen, structures, like flowers.

基因组倍增提供了额外的遗传物质，使植物（或动物）出现新的特征。基因组倍增使植物能够进化出新的前所未有的结构，比如花。

Flowering plants and insect pollinators evolved together. This co-evolution drove plants and insects to diversify unusually rapidly. Flowering plants outcompete rival plants by growing faster and taking up the greatest possible share of nutrients available.

开花植物和传粉昆虫一同进化。这种共同进化促使植物和昆虫异常迅速地多样化。开花植物通过更快速生长和尽可能多地吸收可利用的营养来与其他植物竞争。

Flowering plants have more efficient leaves, with double the number of veins, more efficient at photosynthesis than conifer needles. Conifers do, however, compensate, as all needles on one tree combined, have a larger surface area than broadleaf trees.

与针叶相比，开花植物的叶子功能更强，叶脉数是针叶的两倍，光合作用也更有效。然而，针叶树弥补了这一不足，一棵树上所有针叶加起来的表面积比阔叶树要大。

There are two food webs: the one above the ground powered by the sun, and the underground web, with its energy source made up of decaying organic matter, called detritus. Plants feed herbivores. Detritus from plants and animals feeds bacteria and fungi.

有两个食物网：一个在地面上，由太阳提供能量；另一个在地下，其能量来自腐烂的有机物，称为腐殖质。植物为食草动物提供食物。动植物的腐殖质是细菌和真菌的养料。

Bacteria and fungi are food for mites, springtails, eelworms, pot worms, and earthworms. These species, that are also eating plant roots, are food for birds, and reptiles like salamanders and lizards. Plants and animals feed the world.

细菌和真菌是螨虫、跳虫、线虫和蚯蚓的食物。这些物种也吃植物的根，自身却是鸟类和蜥蜴等爬行动物的食物。植物和动物养活了世界。

Three flower families (sunflower, orchid and legume), with 62,000 different kinds of flower, form about 25% of the world's plant diversity. The smallest flower (Wolffia microscopica) weighs 150 micrograms, and the entire plant is one millimetre long.

向日葵、兰科和豆科这三个花科有62 000种不同的花，占世界植物多样性的25%。世界上最小的花（无根萍）重150微克，整株植物有1毫米长。

Would you like to live under or above the ground?

你愿意住在地下还是住在地上?

What about getting food from decaying plants?

从腐烂的植物中获取食物会如何?

Are flowers important in your life?

花在你的生活中重要吗?

What is your favourite flower?

你最喜欢什么花?

Flowers have many different colours and shapes. Did you know that they have developed these in order to attract different pollinators? Do some research into the different ways in which flowers have evolved to welcome their pollinators. Now list the ten you find most interesting. Compare your list with that of your friends. Look at all the adaptations, and then vote for the three most popular adaptations. Work together to compile more detailed information on your top three. Share your research with others, to illustrate just how diverse, interesting and inspiring the plant kingdom is.

花朵的颜色与形状各不相同，你知道这些是它们为了吸引不同的传粉者而进化来的吗？对花朵的不同进化方式做些调查，看看它们是如何吸引传粉者的。列出你认为最有趣的 10 种，和你朋友的调查做些比较，看看其中有哪些适应性。选出 3 个最常见的例子，大家一起去寻找有关它们的更多详细信息。把你们的研究与他人分享，让他们知道植物王国是多么多样化、有趣和令人惊叹。

学科知识

Academic Knowledge

生物学	被子植物和裸子植物；全球植物分布在6个植物区（泛北极、古热带、新热带、南非、澳大利亚、南极），下属37个植物地区100多个植物省；非洲菊是多年生草本植物；非洲非洲菊属金盏花亚科；被子植物占据植物种类和生态系统中的大多数；基因组倍增改变DNA的技术；开花植物的减数分裂和有丝分裂。
化　学	非洲菊有止血特性；分解：指土壤有机体用酶分解有机质，留下腐殖质，帮助形成土壤；DNA指挥细胞产生各种颜色的色素；蓝色、红色、粉红色和紫色色素都含有花青素，类黄酮的一种，胡萝卜素产生红色和黄色。
物　理	花瓣细胞产生的色素能吸收除红色外所有颜色的光，花朵会反射红光所以呈红色；花朵颜色会随时间而变淡，比如从粉色变成蓝色，这向传粉者发出信号，表明花朵已不再需要传粉。
工程学	电子和机械的连接与紧固设备都是成对的，比如插座和插头，螺母和螺栓。
经济学	粮食生产依赖阳光和土壤，具有极高的适应性和多样性，这与工业和市场营销中统一的标准化方法截然不同。
伦理学	为确保合作顺利而随时准备作出调整。
历　史	被子植物出现于大约1.3亿年前的白垩纪中期；被子植物在中国出现的时间比其他地方早了1 000万年；在罗马时代，外科医生用非洲菊来治疗伤口；1974年，植物学家罗纳德·古德确定了六个植物区系；感恩而死乐队录制了流行歌曲《白玉兰》。
地　理	植物物种组成相对一致的地理区域称为植物群落；植被张力带；在北半球，针叶树主要分布在寒带和温带；玉兰原产于亚洲东部和东南部，以及加勒比海地区和中美洲；白玉兰是上海市花；天女花又叫木兰，是朝鲜国花；南非西海岸国家公园和波斯特堡野花保护区有着最迷人的非洲菊。
数　学	百合有3片花瓣，毛茛有5片，菊苣有21片，非洲菊有34到55片花瓣；向日葵的花盘有两组朝不同方向盘绕的螺旋线，每组螺线的花瓣数不同，比如21和34，34和55，55和89，或者89和144，这些都是斐波那契数！
生活方式	在日本，玉兰的花蕾和嫩叶被当作蔬菜食用；在中国，玉兰花被用作中药材，称为厚朴。
社会学	玉兰象征自由、浪漫和真爱永恒；非洲菊代表着快乐、积极向上、简洁；日本社会对樱花花期的监测；死亡是生命的一部分。
心理学	鲜花对情绪健康有强烈的积极影响；使家人和朋友亲密无间；人们习惯把红花与爱和浪漫联系在一起，红花对免疫系统有促进作用；紫色的花对心情和神经有镇定作用，能使人振奋并激发创造力；黄色的花能带来快乐，刺激记忆和神经系统；白色的花代表纯洁、诚实和完美。
系统论	植物及其花朵能适应不同的传粉者，如蜜蜂、苍蝇、蝴蝶、蜂鸟、蝙蝠或啮齿类动物；授粉生态与植物进化的相互作用，也叫作授粉者转移。

情感智慧

Emotional Intelligence

非洲菊

非洲菊自尊心很强。她担心自己在炫耀美貌，准备好了做自我批评。随着和玉兰的对话轻松展开，她的自信也在增强。她表达了感激之情，并对玉兰的领导力表达敬意。她肯定了昆虫的作用，并再次赞扬开花植物能让昆虫为他们授粉。总的来说她见多识广，但得知腐质碎屑是健康土壤的来源时还是很震惊。她欣然地谈论生态系统中其他伙伴，以及叶子变成花朵的信息。她谦虚地列举了一些其他的花科。她对人类的技术表示欣赏，但同时指出是花朵的独特性促进了生命繁荣。

玉兰花

玉兰花觉得非洲菊最有活力，她们如地毯般成片绽放。他提醒非洲菊昆虫在花的进化中起主导作用。当非洲菊说起他在蜜蜂的授粉中起着关键作用时，他仍然保持谦卑，并赞美蜜蜂的快速学习能力。玉兰指出了土壤和土壤中生物的作用。面对非洲菊的震惊，他耐心解释说死亡是生命的一部分，并详细说明了这个过程。玉兰对自己从叶子变成花朵的独特能力感到自豪。

艺术

The Arts

花呈现出各种惊艳的色彩，插花就是对花朵和颜色的艺术应用，创造一个有趣的插花吧！我们知道颜色代表着不同的情绪或心境，还会对心理产生影响。先调查一下不同颜色的象征意义，然后选择三种不同颜色的花：一种让你联想到你的母亲，一种让你联想到你的父亲，另一种给你自己。用简单而有趣的方式把它们整理一下，然后送给你的父母，并告诉他们每一朵花所代表的意义。

思维拓展

Systems: Making the Connections

地球上的生命都与适应性有关。35万种开花植物已提供强有力证明，只要能够适应就能够生存，并最终促进生命繁荣。开花植物出现较晚，但繁殖很快，这与达尔文的进化论不太相符，直到今天也没能完全被进化生物学家解释清楚。如果没有开花植物，我们就不会有粮食作物，也就不能喂养牲畜。地表的光合作用和地下由腐殖质到肥沃土壤的转化不仅是生产力的基础，也是调节地球健康最重要的碳汇之一。为了实现可持续发展，我们不仅要尽最大努力保护生物多样性，还需要了解联结土壤和生物多样性的复杂系统。标准化消费下的工业化种植与自然的运作方式完全矛盾，这启发我们必须找到更好的商业模式，不是在世界各地制造更多相同的东西，而是从每一个生命网中获得最好和最多的东西。植物与动物的神奇交汇使植物进化出花，为进化创造条件，并满足动植物的基本需求。不断进化应对变化，这是地球生命的一个特征，大自然提供了一个鼓励改变的框架。我们需要在自然和社会之间找到一个接口，使资本人性化，并激励创新以追求韧性。如果我们能成功地做到这一点，并拥有与“花的力量”同样的天赋，那么向可持续社会过渡将比以往想象的更容易。

动手能力

Capacity to Implement

你了解扦插栽培吗？让我们来学习如何扦插。要使花茎生根，你必须先找到切花茎上的节点。你需要找到至少两个节点：一个用来生根，另一个用来发芽长叶。没有节点就不能生长。将底部节点下半厘米和顶部节点上半厘米处剪下。为了减少水分流失，把茎上的叶子都拔掉。将根茎底端浸在生根粉中。生根粉可以保护花朵免受真菌的攻击。你也可以把它含在嘴里！ 这么做确实有用，因为唾液能抵抗真菌。你还可以用苹果醋或肉桂。把植物的茎放在能保持水分且不太密实的培养基中，比如珍珠岩、蛭石、沙子或椰糠。等待并观察，看看植物是如何表现它的生命力的。

故事灵感来自

This Fable Is Inspired by

艾格尼丝·迪林格

Agnes Dellinger

艾格尼丝·迪林格生于奥地利，毕业于维也纳大学，获生态学学士学位。之后在同一所大学获得了社区与景观生态学硕士学位。第二年，她开始在植物与生物多样性系担任研究助理。2019年，她完成了博士论文，主题为传粉者的转移和花的进化。艾格尼丝和她的同事证明了花朵的形状是为了适应特定的传粉者而进化的。艾格尼丝的研究揭示了花是如何改变并创造出非凡的多样性来吸引、回馈和适应不同的传粉者的。

图书在版编目(CIP)数据

冈特生态童书.第七辑：全36册：汉英对照 /
(比)冈特·鲍利著；(哥伦)凯瑟琳娜·巴赫绘；
何家振等译.—上海：上海远东出版社，2020
ISBN 978-7-5476-1671-0

Ⅰ.①冈… Ⅱ.①冈… ②凯… ③何… Ⅲ.①生态
环境-环境保护-儿童读物—汉英 Ⅳ.①X171.1-49

中国版本图书馆CIP数据核字(2020)第236911号

策　　划　张　蓉
责任编辑　程云琦
助理编辑　刘思敏
封面设计　魏　来李　廉

冈特生态童书
花的力量
[比]冈特·鲍利　著
[哥伦]凯瑟琳娜·巴赫　绘
颜莹莹　译